W9-CTS-572

CENTIPEDES

ANIMALS WITHOUT BONES

Jason Cooper

Rourke Publications, Inc.
Vero Beach, Florida 32964

© 1996 Rourke Publications, Inc.

All rights reserved. No part of this book
may be reproduced or utilized in any
form or by any means, electronic or
mechanical including photocopying,
recording or by any information storage
and retrieval system without permission
in writing from the publisher.

PHOTO CREDITS
© Alex Kerstitch: cover; © James P. Rowan: title page,
pages 4, 8, 13, 15, 18, 21; © James H. Robinson: page 7;
© Lynn M. Stone: pages 10, 12; © Barry Mansell: page 17

Library of Congress Cataloging-in-Publication Data
Cooper, Jason, 1942-
    Centipedes / by Jason Cooper.
        p. cm. — (Animals without bones)
    Includes index.
    Summary: A simple introduction to the physical characteristics,
life cycle, and habitat of centipedes and other myriapods.
    ISBN 0-86625-574-5
    1. Centipedes—Juvenile literature. [1. Centipedes.] I. Title.
II. Series: Cooper, Jason, 1942-   Animals without bones.
QL449.5.C66 1996
595.6'2—dc20   95-46037
                                 CIP
                                 AC

Printed in the USA

TABLE OF CONTENTS

Centipedes   5
What Centipedes Look Like   6
Kinds of Centipedes   9
The Centipede Family   11
Where Centipedes Live   14
Baby Centipedes   16
How Centipedes Live   19
Predator and Prey   20
Centipedes and People   22
Glossary   23
Index   24

CENTIPEDES

You can't lose a leg and move very quickly, but a centipede can. In fact, a centipede can lose 10 or 12 legs and still run!

Centipedes can afford to lose a few legs. Each centipede has 30 to 354 legs.

Centipedes look a little like caterpillars. Caterpillars, though, are insects. Centipedes belong to a group of small, boneless animals called **myriapods** (MEER ee uh pahdz).

Centipedes look like insects, but they're myriapods

WHAT CENTIPEDES LOOK LIKE

People often think centipedes have 100 legs. "Centipede" comes from Latin words that mean "100 feet." However, centipedes never have exactly 100 feet. That's because a centipede always has an odd number of leg pairs — like 21, 33, or 47.

A centipede has a thin body that has many sections called **segments** (SEG ments). Except for the first segment and the last two, each segment has a pair of walking legs.

Like sausage links, a centipede's body is made up of segments

KINDS OF CENTIPEDES

Scientists have named about 2,800 different **species** (SPEE sheez), or kinds, of centipedes. They can be from about 1 inch to 1 foot long.

Centipedes are often brown, like dirt and dry leaves. A few kinds are yellow with green stripes.

Centipedes and millipedes are close cousins. Millipedes have two pairs of legs for each body segment. Millipedes can have as many as 752 legs, but most species have about 100.

The house centipede blends easily with the woodwork

THE CENTIPEDE FAMILY

Myriapod means "10,000 feet." Myriapods — centipedes and their cousins — *do* have many legs. None of the myriapods, though, really has as many as 10,000. Imagine trying to count the feet of a myriapod as it scurries past!

Myriapods are related to groups of animals like insects, crabs, and spiders.

The millipede has two pairs of legs on each body segment

A millipede tries to hide in the leaf litter of a Costa Rica forest

With two pairs of legs on a segment, how many legs do these Kentucky millipedes have?

WHERE CENTIPEDES LIVE

Centipedes live on land — wherever they can find food and damp hiding places. You might be living close to a centipede if your home has a basement or attic. Centipedes in woodlands live under rocks and leaves. They also burrow, or dig, into moist soil.

Don't look for centipedes in ponds or rivers. These creatures love dampness, but they can't live in water.

Centipedes like damp earth and the cover of leaves

BABY CENTIPEDES

Some kinds of centipedes lay a pile of eggs. The mother centipede coils, or curls, around the eggs and waits until they hatch. Sometimes she waits three months!

Other kinds of centipedes lay eggs one at a time and cover each egg with dirt. After the eggs have been laid, the mother centipede hurries away.

When centipedes hatch, they look like tiny adult centipedes. Some kinds take a few years to reach adult size.

A mother centipede coils around her eggs

HOW CENTIPEDES LIVE

Centipedes, like most owls and bats, are **nocturnal** (nahk TUR nul) — they are creatures of the night. Nights are usually damper than days, so centipedes like the darkness.

Centipedes move quickly. Their legs all work together. It's interesting that the fastest centipedes are not the ones with the most legs. The fastest have about 15 pairs of legs.

Centipedes are active at night

PREDATOR AND PREY

Centipedes are **predators** (PRED uh torz), or hunters. They snack on other little animals. Small centipedes eat bugs, snails, worms, and other soft, little creatures.

The largest centipedes, those in tropical countries, have big appetites. These centipedes live in hot, humid places and eat mice, birds, lizards, and snakes.

Centipedes seem to find their **prey** (PRAY), or food, by touch. As centipedes run about, sooner or later they bump into a meal by feel.

Is the centipede coming, or going? Its head is at the left

CENTIPEDES AND PEOPLE

Centipedes are best left alone. Their first pair of "legs" is actually a pair of poisonous fangs. Centipedes use these fangs to fight — and bite.

The poison in a centipede's bite is painful, but rarely serious to people. The bite of the foot-long desert centipede of Arizona, however, can be dangerous.

Many people find that little centipedes living in their homes are helpful. Household centipedes eat insects and spiders.

Glossary

myriapod (MEER ee uh pahd) — a group of quick-moving, boneless little animals with 30 or more legs, in pairs, on body sections; millipedes and centipedes

nocturnal (nahk TUR nul) — active at night

predator (PRED uh tor) — an animal that kills other animals for food

prey (PRAY) — an animal that is killed by another animal for food

segment (SEG ment) — a section; one separate piece of a whole body

species (SPEE sheez) — within a group of closely-related animals, one certain kind, such as a *polar* bear

INDEX

caterpillars 5
centipedes
    segments of 6
    sense of touch in 20
    species of 9
crabs 11
eggs 16
fangs 22
home 14
insects 5, 11, 22
legs 5, 6, 9, 11, 19, 22
millipedes 9
myriapods 5, 11
poison 22
predators 20

prey 20
spiders 11, 22
woodlands 14